HELP DEVELOP EARLY MATH SKILLS WITH THIS POETRY BOOK

Help Develop Early Math Skills with this Poetry Book

Counting to 50

Walter the Educator

SKB

Silent King Books a WhichHead Imprint

dedicated to the parents that genuinely care about their kids' education

WHY I CREATED THIS BOOK?

Educationally, research shows that kids will benefit when they develop early math skills; this book intertwines the love of poetry with those early math skills.

Counting to 50

Counting up to fifty, oh what fun!
Let's get started, one by one.

1

**One little birdie up
in the tree**

2

**Two fluffy clouds,
how pretty they be**

3

**Three little kittens,
playing with string**

4

Four buzzing bees,
ready to sting

5

Five colorful flowers, in a row

6

Six busy ants,
on the go

7

**Seven jumping frogs,
in the pond**

8

**Eight sweet
strawberries,
so juicy and fond**

9

**Nine shiny stars,
twinkling bright**

10

**Ten ripe apples,
a delicious sight**

11

Eleven hopping
bunnies,
in the grass

12

Twelve chirping
crickets,
as we pass

13

Thirteen big
balloons,
soaring high

14

**Fourteen
shimmering fish,
swimming by**

15

Fifteen busy bees,
making honey

16

**Sixteen soft pillows,
so comfy and cozy**

17

Seventeen tasty cookies, fresh from the oven

18

Eighteen buzzing
bees,
still a-buzzin'

19

Nineteen jumping
beans,
bouncing with glee

20

**Twenty fluffy clouds,
as far as we can see**

21

Twenty-one little
ladybugs,
crawling on a leaf

22

**Twenty-two shiny
marbles,
what a relief**

23

Twenty-three colorful crayons, so bright

24

**Twenty-four
sparkling diamonds,
a beautiful sight**

25

**Twenty-five
racing cars,
zooming by**

26

**Twenty-six
soaring kites,
up in the sky**

27

Twenty-seven busy
bees, in their hive

28

Twenty-eight ripe bananas, ready to thrive

29

Twenty-nine
bouncing
balls, bouncing
so high

30

Thirty twinkling
stars, up in the sky

31

**Thirty-one pairs
of socks, so neat**

32

**Thirty-two
yummy cupcakes,
a tasty treat**

33

**Thirty-three
friendly dolphins,
swimming along**

34

**Thirty-four
bright rainbows,
after a storm**

35

**Thirty-five
pretty butterflies,
fluttering about**

36

**Thirty-six
fluffy clouds,
without a doubt**

37

Thirty-seven
juicy oranges,
so sweet

38

Thirty-eight
busy ants,
marching to the beat

39

**Thirty-nine
bouncing kangaroos,
hopping around**

40

**Forty sparkling
diamonds,
on the ground**

41

**Forty-one
colorful balloons,
floating in the breeze**

42

**Forty-two
buzzing bees,
flying with ease**

43

Forty-three
playful puppies,
wagging their tails

44

**Forty-four
shiny pennies,
in shiny pails**

45

Forty-five racing rockets, shooting for the stars

46

**Forty-six
twinkling diamonds,
Venus and Mars**

47

**Forty-seven
busy bees,
buzzing all day**

48

**Forty-eight
colorful flowers,
in a big bouquet**

49

**Forty-nine
bouncing bunnies,
so cute**

50

**Fifty shining stars,
making us mute**

ABOUT THE AUTHOR

Walter the Educator is one of the pseudonyms for Walter Anderson. Formally educated in Chemistry, Business, and Education, he is an educator, an author, a diverse entrepreneur, and the son of a disabled war veteran. "Walter the Educator" shares his time between educating and creating. He holds interests and owns several creative projects that entertain, enlighten, enhance, and educate, hoping to inspire and motivate you.

www.ingramcontent.com/pod-product-compliance
Lightning Source LLC
Chambersburg PA
CBHW020343130626
46549CB00003B/1261